Lazarus Revived

Lazarus Revived

AN ATHEIST ARGUMENT FOR CONSCIOUS LIFE AFTER DEATH

Alexander Matthews

FIREFLY BOOKS

Published by Firefly Books Ltd. 2019

The cartoons are brilliantly conceived and drawn by another Alexander Matthews, the cartoonist, and are used by permission.

First printing

Library of Congress Control Number: 2019933684

Library and Archives Canada Cataloguing in Publication
Title: Lazarus revived : an atheist argument for conscious life after death / Alexander Matthews.
Names: Matthews, Alexander, author.
Description: Includes bibliographical references and index.
Identifiers: Canadiana 20190066121 | ISBN 9780228102182 (hardcover)
Subjects: LCSH: Immortality (Philosophy) | LCSH: Consciousness. | LCSH: Death. | LCSH: Cosmology.
Classification: LCC BD421 .M38 2019 | DDC 129—dc23

Published in the United States by
Firefly Books (U.S.) Inc.
P.O. Box 1338, Ellicott Station
Buffalo, New York 14205

Published in Canada by
Firefly Books Ltd.
50 Staples Avenue, Unit 1
Richmond Hill, Ontario L4B 0A7

Cover and interior design: Jacqueline Hope Raynor

Printed in the United States of America

Contents

To my late brother Paul
who was interested in such speculations

1

Do Your Beliefs Make Sense?

Of course there is conscious life after death.
There is a good argument for it.
What do you think?
Surely this matters a good deal to you?

Impossible.
There is no conscious life after death.
There is proof that there is none.
What a let down.

How do you reconcile these two statements? That's the purpose of this little book.

The subject of this book, that there is good scientific reason for conscious life after death, seems so fascinating and important that I can't resist putting down what I think. I hope this book will encourage someone else to take up the cudgels and mull over the idea in a much more esoteric and organised way. As we shall see, conscious life after death is already in the literature. What this book does is to expand on the idea by expanding our sense of what time is. Until we're able to talk of conscious life after death as meaning, for example, that someone is getting younger.

I realise that this is not the way for an ex-scholar to write. But I've been thoroughly seduced by the idea of conscious life after death, particularly since it reverses some of my former beliefs. Yes, I've had a conversion. Not to Christianity, but a conversion to do with science – science and the afterlife. It's a revelation to me that science has something to say about conscious life after death. It should be to you, too.

The vast philosophical literature on consciousness seems very confused. *The Oxford Compact English Dictionary* defines consciousness as "awake and aware of one's surroundings and identity." Let's stick with that for the moment. After all, consciousness may vary greatly from person to person. I may not be aware of what you are concentrating on. We may be both unaware of our surroundings because we are focused on something else. While we are looking through microscopes, our daughters are sneaking off with wild, hairy dope addicts, our wives are making eyes at the milkmen and our sons are having nervous breakdowns. Work has made us oblivious to family needs. Consciousness is, after all, a very personal matter. Having said that consciousness, as I shall try to argue in this intellectual adventure, consists of a series of human made artificial constraints. Made by you and me. These constraints allow us to see in the way that we do, to make connections and even, to some extent, control what we see. Consciousness is not already in the brain, though the capacity for it is. Brain experts look for it there, but it isn't there at all...it is made by us as we grow into adults. It is not a physical something, it is a manufactured series of mental strait jackets we have to acquire in order to see and experience life as we know it. To realise this is to further expand our human consciousness. We shall see how this works as the argument for conscious life after death unfolds. Only when these constraints are fully developed in us can we finally say that we are conscious. It is very odd that in order to expand our consciousness, we have to narrow what we see by a series of constraints. How we impose these restraints on seemingly chaotic and unexplainable experience seems even odder.

Let's look at this oddness more closely. If I say to you that there is a good argument for time reversal, what are we talking about when we say that? This is not a philosophical statement. Philosophy doesn't have a subject matter, no matter how hard it's tried to come up with one. No, an argument for time reversal is a statement about cosmology and in this book I'm going to cling to cosmology like a limpet. Then I can say, that's true or that's false, depending on the scientific implications. To discuss the implications of a statement about time reversal you have to have a subject matter you can be right or wrong about. So cosmology, the theory of the origin of the universe, is the subject matter here. For the purposes of this essay, we are thinking like cosmologists, not philosophers.

Philosophers might jump in at this point and try to usurp the situation by asking us to revisit the vast literature concerning what "consciousness" actually is. How we can know anything? What is it to be ourselves? If we trade brains, who are we?

Those questions, fascinating as they are, are not our subject here. So, sticking to cosmology like a limpet means sticking to cosmology like a limpet. If we're going to talk about cosmology, then we have to take consciousness for granted. We are conscious and we are limiting this intellectual adventure to cosmology.

Let's think for a minute about the obvious objections to what we're talking about here. Could there possibly be good scientific reasons for supposing there was conscious life after death? We know that when we die our brain cells die and that consciousness stops when our brain dies. So, how could conscious life continue after we're physically dead?

How can I argue scientifically for conscious life after death when science seems to show us that the opposite is true? It seems there is no way to argue for life after death scientifically. And yet I seem to be asserting that there is.

There are people who pay for cryonics so that they can be brought back to life in the future. Others suppose they will meet their friends in heaven. It's hard to get it into our heads that we will have lost consciousness at the moment of death, that when the brain dies, that's it. Are you really willing to accept this very hard fact of life?

Yet, despite all this, in this intellectual adventure, I want to argue that there's a good scientific reason for talking about conscious life after death. For the first time in human thought, as a result of the proofs involved with general relativity, this argument now seems possible. So, this should be the most exciting news you've ever heard in your life.

You do not have to be religious to believe in this form of life after death. In fact, my thesis is supported strictly by good scientific reasons. That's very strange because if you're a real scientist, the limits of what you know depend on what you see and what conclusions you can draw from what you see. Admittedly, science is a work in progress. An unfinished sculpture, if you like. Sometimes we have to abandon the sculpture we've grown comfortable with and start over again. We've experienced a complete scientific revolution. Everything you and I thought we knew has flipped into a totally different way of looking at things. This is what happened, for example, when people realised that the Earth went around the Sun rather than, what seemed self-evident at the time, that the Sun went around the Earth.

Do you feel threatened by these assertions about life after your death? Is your belief system in disarray? As scientists, at least, we are open to changing what we think is true. Doing science is like being a mental tourist and travelling to places where no-one's ever been before. Very exciting. The first footfall on mental alien territory. Science does you the favor of expanding your human consciousness. That should be very exciting to you but perhaps your beliefs are challenged by science? If so, you may get very nervous and uptight about all this and insist that what you believe is true. I do too, to some extent. For years I resisted what I'm about to suggest until I was able to make a major flip in my thoughts.

There are a number of points to make before this rather outlandish assertion makes sense. If you get impatient with these points, you can always skip to the thought experiment towards the end of this little book (Chapter 15) and just read that as an introduction and then read the assertion itself. However, for those of you who want to make sense of the whole thing, I suggest that we go through the following considerations together and share the full intellectual adventure surrounding this statement. It may not be a completely comprehensive explanation to you; you'll have to add your own bits to it here and there, but I will be as comprehensive as I possibly can be. To be comprehensive is to be comprehensible and vice versa. So let's begin.

We have to deny you health insurance because of your non-existent condition.

2

Comprehensiveness Is Comprehensible

If I make the assertion that there is conscious life after death, how can I be sure that it's comprehensible and believable unless it meets all objections? Does the assertion itself rest on assumptions or theories that need to be proven? Does it rest on just a limited class of observations, neglecting other observations, other quantities, that would invalidate the assertion?

Suppose you were given the job of describing a piano to people who had been locked up in a room devoid of all sensory or material experience of any kind for their entire lives. How would you go about describing a piano so that they knew what you were talking about? First of all, you'd have to teach them a language so you could communicate. Then you'd have to somehow get across to them what the word "piano" refers to – give the word some semantic content. You'd be starting from scratch. Somewhere in that explanation you might get the idea that you're having to describe to this person your entire experience of the world to get across what a piano really is. This is called the doctrine of holism. The trouble with it is, no one can come up with a complete holism – a complete view of all their experience at once. Let alone all experience.

This process of describing a piano could involve what I would call a double induction. The person would induct the same visual experience you're repeatedly having and then notice your constant reaction to that visual experience. He could then repeat or imitate your reaction and get the idea of simple referential meaning through this double induction. He cannot understand what a piano is until he has had enough experience to contextualise it. This is not a complete holism, but at least a small part of it.

So, in making this assertion that there is conscious life after death, I'm presuming that we all share a context of conventions and theories that allow us to communicate. But whenever you make presumptions, those presumptions can be wrong. If this happens the meaning of what I'm saying goes astray because we don't understand each other and I'm not getting across to you what I mean when I say there is conscious life after death.

As an example, consider the following situation: four men are at a party and decide they will play a little game. The game goes like this: each man will describe the most interesting person he's met in the past week. The winner is the man, they all agree, who has met the most interesting person.

The first man says, "I met someone who was publicly and strongly desired by Marilyn Monroe."

The second man says, "I met someone who was castigated for helping a girl with her mathematical homework by the local teacher. During the reprimand, the teacher noticed that he wasn't wearing socks."

The third man says, "I met someone who refused the offer of being prime minister of Israel."

The fourth man says, "I met someone – top this – who refused his salary because he thought it was six times too large."

Fortunately there is no need to choose which man met the most interesting person because these facts are all ascribable to the same man – Albert Einstein. None of the speakers knew they were talking about the same man. In order to figure that out, they had to name him, show photographs or produce other criteria that would identify the man they were describing.

Here are two further examples that show how your experience and education and the context in which you live can affect what you see. Think of a gourmet, a jockey and a veterinary surgeon looking at the same horse. Think of what they might be thinking. Or perhaps think of a young boy, an astronomer and an astrologer. They are all looking at Betelgeuse.

How can I be sure that the demands of comprehensiveness are met, that you understand the same conventions that I do? This isn't always the case. We speak to each other with confidence, but we're not always understood, perhaps even not as often as we think we are. Just look again at the above examples.

Somehow, in order to communicate accurately when we are talking to others, we have to find ways of defining precisely what we refer to by the words we use. This is where science excels with its mathematical equations, laws and definitions. Science makes it possible to precisely define the content of our experience. In order to do this, science relies on thinking and experiencing the world with a theory-ladenness – using theories that do not exist in nature but which create an organised picture or assumption we can comprehend. Think of how we use the theory of hours and minutes to measure time. Think of how we use the equator to neatly divide the globe.

One thing is certain, though. If, in this book, I make a claim that sounds outlandish, which goes against what science seems to tell us and if, at the same time, I'm using science to make the claim, I'd better be very sure of what I'm trying to tell you. And that, folks, is not easy.

We all look at problems and issues from different perspectives. If you visualise a problem, think of it as a mountain. Your explanation and my explanation may be different ways of going up this mountain. An explanation can give very little sense of the true size and shape of the entire mountain if it is only focused on one path up to the top. People approaching, let's say, Mount Everest from very different directions will seem to see very different mountains. So, how do you identify this mountain, let alone give some notion of its true grandeur and size? The best writers try to be comprehensive. Wittgenstein was like a Rottweiler. He used to go around a problem making what looked like unrelated remarks about it. But they were related in the sense that they were about the same problem.

Stephen Hawking made a very salutary comment about this in his book *A Brief History of Time*. Hawking says: "It turns out to be very difficult to devise a theory to describe the universe all in one go. Instead we break the problem up into bits and invent a number of partial theories. Each of these partial theories describes and predicts a certain limited class of observations neglecting the effects of other quantities or representing them by simple sets of numbers. It may be that this approach is completely wrong. If everything in the universe depends on everything else in a fundamental way it might be impossible to get close to a full solution by investigating parts of the problem in isolation… ." [1]

All we can do is try to refine the particular idea we're talking about, in this case, conscious life after death, as best we can. We do this with criteria to do with measurement and identification.

3

Hypotheses and Thought Experiments

Another problem for clear communication is that we may not be able to see directly the ideas and objects that we're talking about. In this case, our epistemology—our theory about what it is possible to know—is not established solely by direct observation. For example, it may be that we will never be able to develop a powerful enough telescope to see as far as a distant light source. But suppose, hypothetically, that you could develop such a telescope and see the distant light source. Just as travelling on a beam of light for Einstein was hypothetical.

This kind of "what if" thinking is what we mean by a thought experiment. It's like a leap of the imagination. Your leap of the imagination gives you an insight into what you would see supposing we were powerful enough to, in this case, make the observation. Admittedly, knowledge gained from these thought experiments is less certain than knowledge gained from actual observation. On the other hand, they give us further insights that are available in no other way into the secrets of what might someday become a science – the screaming secrets.

Again, there are those who doubt the value of thinking itself and who suppose that "there are more things on heaven and earth, Horatio, than are dreamt of in your philosophy."[2] I can't tell you the number of times people have quoted that line from Shakespeare to me and then given me a sad look, as if to say, "You poor dear, you're stuck in rational thought as though it were the be-all and end-all of your existence. And you're cutting yourself off from human relations." As though to believe in some kind of "supernature," as well as our scientific explanations of nature as all that we can know, meant that you were open to more knowledge, more experience, more emotions and more feelings. To rational thinkers that seems nonsense.

Philosophers, on the other hand, give a sad look to Christians and other believers who say we can believe in things that we cannot prove are true. That, to a philosopher, does not make sense, so there is a standoff here between the Christians on the one hand and the scientists and the philosophers on the other hand. Each camp thinks the other has a screw loose somewhere.

Note that Hamlet's reply to Horatio comes when he's thinking about his father's ghost. That is, he's thinking about the afterlife. It's the afterlife, say philosophers and scientists, that tempts or persuades people to think that there is a supernature that affects their life, as a part of their existence.

If we could show by scientific means, that is from an atheist's standpoint, that there is conscious life after death, would that take away at least some of the ground or some of the temptation to think of supernature as part of our lives, part of our experience, part of our feelings? Maybe it wouldn't, who knows? We'll know more when we get towards the end of this intellectual adventure.

Certainly most philosophers and scientists see contradictions between science and faith while religious people do not. Probably there are more questions than answers in physics. Those of a religious bent would say, "You see, you see, philosophy and science don't know all the answers." And then they would quote Hamlet again. So the debate goes on. In any case, it's your beliefs that we're talking about here.

4

The Triangles of Epistemology

Epistemology for a philosopher simply means the theory of knowledge. Think of our various branches of epistemology as self-contained triangles.

I am approaching the assertion that there is conscious life after death from within a cosmological triangle. That doesn't preclude you from approaching it from within other triangles. It's just that this is the way I'm choosing to approach this subject. It seems a good choice to me because it works with cosmological or astronomical observations and sense experience. It's the ultimate anchor, if you like. It pins things down. If you can't observe it, or the observations clash, then something's wrong with the epistemology and you go back to the drawing board.

Epistemology is a work in progress. Scientists and philosophers regard it as incomplete and continue to try to get a consistent understanding and theory of what's going on around us in the universe. So that means that our knowledge can change and, on some occasions, do a complete flip-flop.

To be clear, we use measurements wherever possible but especially in places where our senses can otherwise trick us. And this, again, is because of the theory-ladenness of our experience of the world around us. For example, if we put one hand on a block of ice and another hand on a heater, and then put both hands in tepid water, we get different messages about the temperature of the water. To the hand that was on the block of ice, the water seems warm. To the hand that was on the heater, the water seems cold. So we need measurements to cover anomalies like that when our sensory experience cannot be relied on.

Here is the way these triangles would work. When you ask a question in the triangle of physics, you get an answer in the triangle of physics. When you ask a question in the triangle of biology you get an answer in the triangle of biology. When you ask a question in the triangle of geography you get an answer in the triangle of geography. And so on. Science helps us understand and even control our experiences. It is when we ask "large questions," questions that do not in and of themselves belong to any epistemological triangle, such as "is there conscious life after death?" that we may seem to come unstuck and find ourselves in scientific limbo. We have suddenly no authority to insist that any one set of beliefs is "true". Yes our conscious understanding seems to be expanding at least in relation to what we think we already know, but "know" here is a dangerous word. As Rovelli put it: "science seems to be a work in progress." The best we can do is to artificially attach our question to a particular epistemological triangle. For the purposes of this little book we will be surrogate cosmologists. This is awkward because our question supersedes cosmology. So this is yet another sense of theory-ladenness. It seems we may have to distort the question somewhat in order to answer it

This means that we have a subject matter – a triangle if you wish – with its own assumptions and rules that give us a way to say whether an assertion is right or wrong. As we shall see later, it is not quite so simple as that. But, again, this involves another layer of theory-ladenness – the distortion of experience. Those triangles are not in and do not come from our experience. We put them there to help us make sense of the world around us. Just as we put, for example, the equator around the middle of the earth. It's real enough. We cross it or we don't cross it. It gets hotter as we get closer to it and so on. But the equator is a man-made invention to measure where we are. It's not something that actually exists in nature. Similarly with these triangles. For all their usefulness, they don't exist in nature.

The awkwardness of having separate subjects and creating an epistemology from within these separate subjects, is that our experience of the world comes to us as all in one piece, all interconnected. And these triangles aren't interconnected. Think of that quotation from Hawking. Some scholars try to do interdisciplinary studies – studying more than one triangle – but, as yet, they haven't got very far. You can regard some of these triangles as touching or interconnected, if you like, but they are, as I see it, separate subjects or self-contained triangles.

One thing is certain. In order to make a scientific statement that can be proven to be right or wrong, we have to do a lot of conventionalizing.

This is because a lot of our observations and what is based on them are theory-laden in several of the senses that we have discovered. No less a physicist than Einstein said, "Insofar as geometry is concerned it says nothing about the outside world, and insofar as it says something about our experience it is uncertain."[3]

So what are we to do about these limits to our knowledge? The fact is that we can only be precise about bits and pieces of what experience is like. Then we hope to fit those bits and pieces together into some sort of a universal theory. That's the way we behave in writing about physics and cosmology. Since this piecing together of various small bits of knowledge into a greater whole is based on their separate conventions, what we're going to end up with is a very, very conventionalised, or as the philosophers call it, theory-laden, assertion. In other words, the understandability or comprehensiveness or comprehensibility of what we're talking about, as though it were a unified whole, is not in nature, it's made up by us.

When you observe something or I observe something, we observe it with these conventions or theory-ladenness in mind. So a simple observation turns out to be not so simple. Also, it's a distortion, if not a downright lie, to separate our experience into these triangles and then join them again into some composite that we call "epistemology." Yet this seems to be what we have to do in order to understand experience in the ways that we do.

Einstein says two important things about this theory-ladenness: "Physical concepts are free creations of the human mind and are not, however it may seem, uniquely determined by the external world."[4] And, "The eternal mystery of the world is its comprehensibility...the fact that it is comprehensible is a miracle."[5]

Compare these statements with that made by Hawking quoted earlier: "It turns out to be very difficult to devise a theory to describe the universe all in one go." Oxford mathematician Roger Penrose says, "The laws of physics produce complex systems and these complex systems lead to consciousness which then produces mathematics which can then encode in a succinct and inspiring way the very underlying laws of physics which give rise to it."[6]

So you cannot say that our observations are a direct experience of what's there. They're a kind of an indirect experience brought to us by conventions. Sometimes these conventions are mathematical, particularly when we have to respond to and communicate with others.

Hawking's claim that "if we do discover a complete theory….we would know the mind of God" sounds a bit too ambitious in relation to all this theory-ladenness.[7] Einstein's more cautious approach seems more in tune with what we have been talking about in this intellectual adventure.

5

Vocabulary Under Strain

As we discussed, theory-ladenness causes us to look at experience from very different perspectives depending on where we live and our cultural background – perspectives that our vocabulary is unprepared to deal with. Different languages seem to split experience up in different ways. These differences are not so obvious when you consider languages that are in the same language family, for example, Spanish and French or English and German. But when you start translating a description of what we experience from, say, English to Swahili or to a Native American language, then you can run up against very different ways of looking at what otherwise might be called the same experience.

For example, suppose you were flying in a plane over the very northern part of Canada where everything is covered with snow and ice and the plane is forced to make an emergency landing. You might survive if you can communicate in the Inuit language but suppose all you have are words in the English language. Your chance of survival is better if you know the native language because the Inuit have a number of different words, maybe seventeen or more, for what we, in English, simply call "snow." They would be able to tell you which snow could be used to build igloos, which snow to avoid so as not to fall through and so forth, whereas we couldn't make those kinds of specific distinctions in the English language. We wouldn't know where to look. Or how to look. Our word "snow" is a blanket term. We have little use for it while, to the Inuit, it is a crucially important term with many subtle distinctions. So their experience has enriched their language in ways that we don't have and don't need, at least not unless we crash in northern parts of Canada.

In the same way, certain African languages simply don't have a word for "brain." The nearest term they seem to have is "ancestor." To eat someone is to assimilate their ancestors. It's a compliment to be eaten.

The Greeks have five words for "love," but the single word "love" in English tends to be somewhat overloaded. It stretches all the way from Johnny Mathis to Mother Teresa and back again. Or, more accurately, what Johnny Mathis sang about and what Mother Teresa did in her lifetime.

So, when we say that there is conscious life after death, we are making that statement within the bias of the English language. I was told that Einstein, when he was speaking about his youth, said that it was an advantage to him that he didn't learn to speak as quickly as other children. He was free of the prejudice of language and could imagine things visually without being tied down to a certain way of seeing things imposed on him by his vocabulary.

Philosopher Willard Van Orman Quine points out that if two lexicographers, language experts, from different languages came upon a very specific experience, let us say a "rabbit", there is not enough sense data available to us to determine how the two lexicographers are going to describe the experience.[8] The sense data alone, the rabbit, doesn't dictate what is seen. What also dictates the experience is the language that is used to describe it.

It's the way each language slices experience that decides what a "rabbit" is, not only the sense experience itself. This is an odd thing to say because we've just said that we're relying on sense experience for epistemology. Yes, we are, but we're relying on it from the point of view of a specific language. So we're stuck with the prejudice of the subject (cosmology), the prejudice of the theory within that subject and the prejudice of the language that we use to articulate that theory. There are problems of identity here, but we won't go into them. There is a vast and sophisticated body of literature on the subject of identity in philosophy but it's beyond the scope of this short intellectual adventure to go into it here.[9]

6

The Anthropic Principle

There is yet another prejudice we have to consider and that's the anthropic principle.[10] The anthropic principle figures largely in our prejudging what we see. There are many ways of stating this principle and there are many ways of talking about it, but the way that's most useful for us is to suppose that we see the things we do because of the way we're built, the way we are. If, instead of being a person, we were, let's say, a conscious proton in a specific atom, we might see experience - our reality - very differently. We might even go so far as to say that reality is species oriented. Different species experience life in such different ways that we could call these variances different realities. If we were a different species, would we be looking at a different universe, or just a different form of reality?

If the anthropic principle means that we're seeing a very slanted or prejudiced form of reality, what does this do to the term "universe?" The term "universe" is supposedly defined as everything there is, "all existing things."[11] But this could be very different from everything we see. And yet epistemology says that the limits of what we see are the limits of what there is in terms of what we know. Some rationalist philosophers would disagree, but this is the definition we will use. The rest is hearsay. For example, humans can't hear certain noises. They are pitched too high or too low for our ears to pick up. We have to use machines to identify them. We know that there are some animals that can hear them better than we can. Our eyesight can't take in the entire colour spectrum either. Do these physical limitations prejudice our understanding of what exists in the world? We can use instruments to increase our sight and hearing, but is that enough? So it's not just that language and theory provide constraints or conventions on what we see and know, the anthropic principle does as well.

So, does a very different theory or way of looking at things produce a very different universe?

The most we can say here is that it produces a very different description of what is around us. How is my contention that there is conscious life after death affected by this? These are the sort of considerations that encourage us to talk about "our" reality or "a" reality rather than "reality."

So epistemology is really rather prejudicial to the language we use, the theories we use, the anthropic principle, where we are in the universe, and so on. In short, our experience of the universe, as we still call it, is theory-laden. And if we keep expanding epistemology – that is, our knowledge keeps growing and we're expanding human consciousness, which is the purpose of science – are we actually creating or furthering a distortion? In other words, are we getting further away from what there is by learning more? In addition, you could be living so far away from me that the edge of everything you can experience is different from the edge of everything I can experience. Therefore, the more we learn, the more our experiences differ and the more our epistemology differs. Is this not a puzzle? Can we even ask and answer the question "is there conscious life after death?" What further prejudices do we have to identify and understand in order to ask and answer this question?

To go back to the idea of a topic or an idea that has the metaphorical shape of a mountain, is that a problem for our intellectual adventure? Again, using the piano image, can we ever explain it fully? Can we ever explain the mountain fully? I doubt we can. So what we're getting across to each other is some sort of meaning that's incomplete, but it's better than nothing. We have to rely on the other person's conventions being about the same as ours if we hope to make ourselves understood. We must make do with that, although scientific measurements can, in some cases, help.

To get anywhere at all, we have to assume that each other's statements literally express what we want to say. This assumption means that yet another form of theory-ladenness creeps into our explanation, into our assertion. There are many instances when we're not being literal when we speak. When a wife tells her husband who has drunk too much and has a very pretty blond lady sitting on his lap at a party, "It's time to go home, dear," she means a great deal more than it is a certain time on the clock. Or, if two men are standing on a station platform and one of the men says to the other, so that his friend's ten-year-old son will not understand, "The train is about to go," what he means is "apologise to your wife before it's too late." The husband and his wife have had an argument and she is leaving unless he apologises. As their best friend, he hopes that the husband and wife will stay together and is urging the husband to speak while there is still time.

These are just two instances where we're not using words literally. It may be that more often than not we're not being very literal when we speak to each other. If we assume that we're being literal when we make such statements as "there's conscious life after death," that assumption brings with it another form of theory-ladenness.

I am you from another reality and seeing you has really cheered me up.

7

A Finite Universe or Pluriverse

Now suppose, as a youngster, you come out of a store and you see a deep, cloudless blue sky and wonder, "Does the sky go on forever? Is it infinite?" On an evening when the night sky is clear and you can see the Milky Way, you might wonder the same thing. "Is the sky infinite, does it go on forever?" If you're an adult and know something of modern astronomy, you might wonder if the universe is subject to infinite expansion.

This kind of wonder is a challenge to epistemology because the theory of knowledge suggests that to know something is to know its beginning and its end, to know it completely. So the concept of infinity, which has no beginning or end, is a major problem for epistemology.

Somehow, to know the universe in the sense of knowing defined by our epistemology, we have to limit the size of the universe. It cannot be "infinite." Here is a traditional way of doing just that. As Carlo Rovelli puts it, "On the surface of the earth, if I were to keep walking in a straight line, I would not advance ad infinitum: I would eventually get back to the point where I started from."[12]

Travel in our universe could work in the same way. If I leave the Earth in a spacecraft and travel in the same direction all the time, I will fly across or around the universe and eventually end up back on Earth. In this sense you cannot travel beyond the farthest star because the light that emanates from the star travels in curves (or geodesics) and it curves in upon itself. Since light is the messenger of the universe (that is, you cannot observe anything that isn't carried to your eye by light in contexts where you are observing the "universe"), any message light does not bring to you is outside experience and therefore outside epistemology. You can talk about things you can't see in hearsay terms, but it's not, strictly speaking, full-scale knowledge. So the sky is not infinite in that sense.

We get rid of the scientific problem of the infinity of the sky by travelling at the speed of light far enough so that you end up where you started and that's the size of the universe, at least in terms of its diametre.

How does this bending of light take place? We know that light, according to the theory of general relativity, actually bends as it travels through space. To understand this idea, think of a trampoline with a large, heavy cannonball resting in the middle of it. Because the cannon ball is heavy, the trampoline sags in the middle. Now, if you are on one side of the trampoline and I am on the other and you roll a tennis ball towards me, the ball will travel around the cannonball to get to me. It will travel in a curve. That's what happens to light when it travels through space. It is affected by the gravity from objects and travels in curves around them. Some scientists refer to this phenomenon as "slowing down light." The amount of slowing down, or curvature, is called the geodesic.

As Clifford Will says in his book *Was Einstein Right?*, "Whether or not the observer uses the words 'light slows down near the Sun' is purely a question of semantics. All an observer of light travelling near the Sun can say with certainty is that he observed a time delay and the duration of the delay depended on how close the light ray came to the Sun. The only sense in which it can be said that light slowed down is mathematical: 'in a particular mathematical representation of the equations that describe the motion of the light ray…the light appears to have a variable speed.'"[13]

Here then is another form of theory-ladenness to do with making the universe finite enough so that we can do the thought experiments that lead to the assertion I am making that there is conscious life after death. We need this theory-ladenness to make the assertion. Look at how many forms of theory-ladenness we've seen so far – and we're not done yet.

Here are two more sets of theory-ladenness, or conventionalizations, of experience or, if you like, distortions of experience. We really have to distort experience in order to explain it and that's a lesson in itself.

First, to return to the example where you live so far away from me that the limits of what you can see do not correspond to the limits of what I can see. In other words, we're each looking at a different universe, or perhaps we should call them pluriverses. Your pluriverse is different from my pluriverse. What then for epistemology? We would have to say that your knowledge is different from my knowledge. We could compare our versions presumably by measuring time and place and so forth. But then, if we accept another's account of their pluriverse, we're stretching what we call epistemology to include hearsay. Or we're saying to ourselves "I don't really know his version and he doesn't really know mine." But if the universe is supposedly everything there is – as opposed to everything we can observe – then "universe" doesn't seem an appropriate term for what we're faced with here. "Pluriverses" seems the better term. "Universe" overreaches what we want to call epistemology, and "pluriverse" does not.

Once again, our grasp of this situation means that we have to prescribe the meaning of what we're talking about or find new words to use. The vocabulary that we're used to, the conventions that we're used to, have to stretch a bit. And that stretching is the expansion of human consciousness that I'm talking about. It is very exciting. Or it should be.

Even if we could somehow reconcile our pluriverses into one overall idea of a universe, there would still be the distinction between talking of the universe as all there is and talking of the universe as all we can be epistemologically aware of. The anthropic principle sees to that. Even the supposed fact that the universe is infinitely expanding, and expanding at greater rates, challenges epistemology itself and challenges, therefore, the very empirical notions on which we base our theory of knowledge. We can't be empirically aware of infinity, for example. So perhaps, in that way as well, talk of pluriverses makes better sense. Think of empirical truth roughly as sense experience or at least based on it.

There may be as many pluriverses as there are perceivers. You could ask the question, "Are there pluriverses without perceivers?" The short answer to that one is no. In other words, it takes a perceiver to create a pluriverse and to that extent, we're all gods. Of course, you can accept hearsay and believe what other people tell you about their pluriverse, but that's somewhat outside epistemology as we're using the term here. On the other hand we can accept hearsay, as we did above and widen our understanding of epistemology to accept such epistemologically loose talk.

We have to get away from uncertainties as much as possible in order to make the statement that there is conscious life after death. In order to do that, we have to conventionalise as best we can. For example, with different pluriverses, you might say that for something to exist we must be intersubjectively aware of it. And in order for that to be the case, we have to conventionalise it and make it theory-laden in ways that we have just discussed. So existence turns out to be a product of a number of conventions to do with language, theory, the anthropic principle and so forth. It may also have to do with hearsay.

8

Mathematics and Physics

Mathematics contributes to the distortions of our understanding of what we see. It makes sense of things by forming relationships between them through equations which help us to understand those things. But equations are not found in nature. Experience may entice us to use equations to understand what we're supposedly seeing, but that understanding is distorted further by these mathematical relationships.

Jaroslaw Mrożek says that mathematics is kind of prismatic.[14] In other words, imagine a prism where white light is separated as it goes through the prism and turns into a whole spectrum of colour. That's the visual image of the distortion Mrożek is writing about.

Mrożek also says that mathematics selects its cognitive content, what can't be understood by mathematics is simply eliminated from it. Presumably, he means concepts like black holes or the Big Bang, for example. Yet we use math to create all sorts of things – television sets, airplanes, all technology. We use the quantum theory to create iPhones and iPads.

But suppose all this knowledge and control over our environment is made possible because of the conditions we've been talking about. Conditions and conventions that aren't in our experience of the world, yet we've put them there. The great mystery is that, even according to these conventions, they seem to work. Talk of pluriverses, on the other hand, is to some extent stepping outside some of these distortions for they cease to apply in these situations. It's funny, in the sense of odd, to localise our control of experience in this way, but that's the way these conditions work.

So, there are certain aspects of mathematics that do distort experience in order to explain experience, and that distortion disqualifies its use in certain contexts that are outside the conditions it's operating on. Those distortions have to be understood in order for us to understand our observations for what they really are.

For example, suppose we learn more through better instruments and technology and our epistemology changes, perhaps even does a flip-flop, as general relativity was a flip-flop from what Newton had to say. Mathematics doesn't do that. It doesn't grow or change in the sense of believing one thing today and something else tomorrow when we learn more. Mrożek notes that mathematics isn't affected by this expanding nature of epistemology[15]. So there are three senses that it's distorting: the prism effect, the selective effect, and the fact that it's not reflecting the growth of cosmology. There may be disputes about this from those who say that mathematics can be reapplied and so forth. William M. Honig thinks that mathematics has a stultifying effect on developments in physics.[16] This backs up what Mrożek says.

There are those who work in physics who suppose that mathematics is the language of physics, that physics can only be understood through mathematics. I'm not sure Einstein would agree with that. He visualised what it would be like to travel on a beam of light. He visualised the problems of gravity by thinking of himself in an elevator. I think other scientists have probably used visual effects to try to understand the world around them.

Yes, mathematics is a great teacher, but only through relationships. Relationships that may not be necessarily there in nature. You get some very different approaches when comparing Einstein on the one hand with Hawking and Penrose on the other. We've already discussed these different approaches and shown that there is no single clear authoritative view that everybody shares about the position of mathematics in relation to physics.

The question remains, if mathematics is so coherent, how do we, as Einstein asks, make sense of a chaotic universe or pluriverse? Is the universe or pluriverse chaotic? This puzzle haunts the application of mathematics to physics and the very nature of physics in general.

9

Now and Actual

There is another issue that we must resolve (when will they ever stop?). As I have said, light acts as the messenger of the universe. It's the postman, if you like, who gives us the news of what's out there. But when we observe, we observe in terms of what we think of as "now" or what is actually happening "now." But what or when is "now" and what is "actual" are very subjective, very ill-defined terms. So it seems that epistemology depends on a very ill-conceived or subjective understanding of what is "actual" or "now." This means that we could share the same theory, the same language and still have very different ideas about what is observable at the moment we say we observe it. You could take a longer look than I and see more and say that longer look is "now." Or you could be in a hurry, look quickly and miss what I've seen. We can't have this subjective experience at the very basis of what is an intersubjective science where there is a correct and an incorrect answer to questions. It doesn't make sense.

Clocks help us to coordinate what "now" or "actual" is but we don't just have to coordinate what we mean when we say "now" or "actual." What we have to do is to restrict it as well. For example, suppose it's 2:14 p.m. on a clock located on Andromeda. We see that it's 2:14 on the Andromeda clock from light signals arriving on Earth from Andromeda. And we see that it's 2:14 on our clock on Earth. But the Andromeda 2:14 light signal has taken two million years to get to us so, again, we can't use this composite "now" in any clear way. It involves many life-spans and many sets of consciousness and therefore many different perceivers of pluriverses. We'd get in a mess if we talked about "now" without taking all this into account. However, this aspect of "now" has its uses, as we shall see later.

When we make an induction we have to be very clear about what the induction is confined to. We are used to making inductions about the earth-bound phenomena that form the basis of our experience. For example, we experience time as going from past to future. We have what we call facts: water boils at one hundred degrees centigrade and sodium and chlorine come together under certain conditions to make salt. These are scientific facts. But they are localised facts. The science on which they are based wouldn't necessarily work in other circumstances, in other parts of our pluriverse.

So localizing our inductions, or observations, to some extent again distorts them. It's the same as when we use mathematics, a specific theory or language or the anthropic principle, we have to distort in order to see. That sounds really odd but that's the way of it. That's one of the most surprising facts of science. Nobody really knows how we make order out of chaos – if nature is in fact chaotic. All of this is terribly up in the air.

Don't worry kid, what he doesn't know is that your consciousness will live on from a certain cosmological perspective.

10

Distorting our Vocabulary

Even within the triangle of cosmology, and the triangle itself is a distortion, we have two theories that don't make much sense together: the quantum theory and the theory of general relativity. Also, we've had to conventionalise infinity down to something that's finite so that we can understand what are we talking about when we talk about the universe or pluriverse. We've had to be very prescriptive in our vocabulary of words like "reality," "universe," "infinity" and "time". Reality may be just our reality, courtesy of the anthropic principle; the universe may be pluriverses; infinity we've tried to do without; and time may ultimately only be understandable in relation to space-time. We have to restrict the open-endedness in our descriptions to observe and to make sense of the statement "the existence of conscious life after death."

This is an important lesson that we must never forget: we distort in many, many ways in order to observe. You may talk about the universe in a traditional way, as everything that is, but perhaps some of the universe will not be observable to you. And I say "perhaps" because we are on very shaky ground here. The part of the universe we can't see is outside of our epistemology, therefore it's unknown. So we can't make solid statements about it. On the other hand, if we redefine what we mean by universe, as everything we can see, and that turns out to be rather different from everything you can see, supposing we're very, very far apart, then we're talking pluriverses. But are we sticking to a meaning of epistemology as everything that can be known? Or everything we can know? In the latter sense epistemology becomes irredeemably subjective.

The question is: Can you still call what we've distorted (in order to understand it) "correct" or "incorrect?" The answer is yes, as long as we realise what we're being correct or incorrect about. We are being right or wrong about a distorted observation. We have to find measureable ways of talking about it intersubjectively – as between us.

11

The Uncertainty of Knowing

A good epistemologist would say if you can't see something, it may not be there. So that means that pluriverses may not be there, epistemologically speaking, if there are no perceivers who are perceiving them.

Now that sounds very odd. It goes against a remark commonly attributed to Bishop George Berkeley that you can know that a tree is there even though nobody is looking at it because you know that God is looking at it. Epistemology says 'no' to that line of reasoning.

If you're unconvinced that things might not exist when they are not under observation, there is a further example, that may shake or increase your faith in the counterintuitive nature of epistemology. Suppose you are in a park on a very dark night. The night is so dark that you can barely see your hand in front of your face. There are some street lamps on the other side of the park that provide the only light. Suddenly you notice a man walking under one of the street lamps. Then you see him walk under another street lamp to the left. There are several more lamps which you may presume he will walk under if he continues in the same direction. But suppose he doesn't. He changes his mind, left something at home perhaps, or misses his girlfriend and turns around. In which case you have taken your inductions, and with it your epistemology, too far. It no longer serves you as knowledge. You cannot be sure, in other words, that the man is going to appear under the street lamps as you thought he was going to.

If you don't see something and experience it directly, then you cannot rely on it as knowledge. But what then is knowledge itself? What have we done to it? We've twisted it into all sorts of contortions in order to make a coherent observation that other people will understand. This twisted view of experience is what we call knowledge. So our theory of knowledge is a tattered thing. "A tattered coat upon a stick. Unless soul clap its hands and sing, and louder sing for every tatter in its mortal dress."[17]

The very assertion that started this adventure is based on a hypothetical situation. What we have to do is to make a thought experiment so that, hypothetically, the statement that "there is conscious life after death" would be valid. This hypothetical nature is admittedly beyond epistemology as we know it. We'd be asserting that something would happen based on what does happen.

That sort of thinking, which led to metaphysics in philosophy, has a very poor history. The philosophical wayside is littered with wrecks that tried to tell us about what "must be" so because of what "is" so. Is this just another wreck? Well, let's just see. Let's just see.

So, is knowledge not quite what you thought it was? Has it received some low blows, during this intellectual adventure? We're now at the point of bringing up the question of existence itself. The definition of the universe according to the *Compact Oxford English Dictionary* is "all existing things." This, in a sense, begs the question of ontological status. What are all existing things? And are they the same for all of us? For a thing to exist, for example, do we have to be intersubjectively aware of it? Are scientific measurements good enough by themselves without human awareness to guarantee existence? Perhaps in some cases but not in others. Do existing things have to be measurable? Is existence dependent, as a number of philosophers seem to say, on a kind of holism (that is, it has to be contextualised into our experience in some way)? Perhaps experience is so subjective that we have to conventionalise it, in some cases very thoroughly, in order to talk of it and make sense to others. Yet another type of theory ladenness.

12

The Big Bang

Again, the Big Bang itself is counterintuitive. It's an affront to the way we do science. The way we do science is through induction and induction usually involves cause and effect. A cause-and-effect chain goes on forever, which actually also defies epistemology and means that we can't know the beginning and the end of a cause-and-effect chain. But neither can we know the Big Bang as the first cause because then it would not be part of the causal chain.

Philosophers and scientists have muttered about what might have taken place before the Big Bang, but nothing very clear has emerged. Scientists triumphantly announce that there is an echo of the Big Bang. Well, yes. But it could be an echo of a Big Bang, a Big Bang that surrounds us, but not necessarily the Big Bang. And there are many,

many reasons why there is a problem with talking about the Big Bang as though it was a scientific fact.[18]

So, what all of this convention or distortion makes us realise is that we need different ways of looking at our experience and that our vocabulary is basically unprepared and ill equipped for this kind of mental work.

What we call light has very different characteristics when it is used in quantum theory than when it is used in discussions about general relativity. If time's arrow works in both directions – that is, from past to future and future to past – then it suggests a kind of free fall with no overall beginning or ending. And if light goes around the event horizon[19] of a black hole, we have to reassess what we mean by time. And also, truth conditions, when they apply to our experience, are usually applied to only one direction of time. We would have to alter those conditions, too.

So all of this means reassessing our vocabulary and prescribing new and more exact meanings for words that fit what science is now teaching us, which is a much wider, much bigger view of what experience is, than we had when we began this adventure together.

Can we talk about a Big Bang but not necessarily the Big Bang? Could the echo of the Big Bang be just an echo of a Big Bang if it surrounds us sufficiently? Would there be such a thing as the Big Bang if there were time reversal in the universe? The two ideas, as we shall see, clash. If in some instances time were reversed and the universe was getting younger, or parts of it were getting younger, then there couldn't be an overall Big Bang. And to talk about a Big Bang means that we would probably be talking about a pluriverse, not a universe. Terminology is up for grabs here. In order to talk of these matters, we might have to step out of accepted conventions and accepted meanings for the words we use. This means redefinitions on a large scale.

13

Are There No Rules of the Universe?

We have yet another form of theory-ladenness or distortion to deal with now. When you talk about the speed of something you talk about it in relation to the speed of something else. For example, if Usain Bolt runs one hundred metres in nine-point-something seconds, he does it in relation to the movement of the Earth around the Sun, which is, I don't know, eighteen miles per second, let's say. But the Earth is moving in relation to the Sun and the Sun is moving in relation to the galaxy. And the galaxy is moving in relation to other galaxies, and so on. So speed, or the time that is taken to achieve this speed, or even space-time, could be regarded as a relation between specific sets of coordinates. And that's a theory-ladenness or distortion that can be added to the others we've discussed.

What we're saying and talking about in relation to this theory-ladenness or distortion is basically that there are no rules of the universe that we can understand as such. There are rules involving a theory of the universe within a specific language and according to the anthropic principle and other restrictions that we have talked about, but that's quite different from having a set of universal rules. We just have to learn to call these distortions experience, but we must recognise them for what they are. Otherwise, with Stephen Hawking, we will think we could come to know the mind of God.[20]

The fact is that, as Rovelli says. "Science is sometimes criticized for pretending to explain everything, for thinking that it has an answer to every question. It's a curious accusation...doing science means coming up hard against the limits of your ignorance on a daily basis....A scientist is someone who lives immersed in the awareness of our deep ignorance....Science is not reliable because it provides certainty. It is reliable because it provides us with the best answers we have at present. Science is the most we know so far about the problems confronting us. It is precisely its openness, the fact that it constantly calls current knowledge into question, that guarantees the answers it offers are the best so far available: if you find better answers, these new answers become science....The answers given by science, then, are not reliable because they are definitive. They are reliable because they are not definitive....It is the awareness of our ignorance that gives science its reliability...the search for knowledge is not nourished by certainty: it is nourished by a radical distrust in certainty."[21] Rovelli goes on to say, "... and this means not giving credence to those who say they are in possession of the truth."[22] According to this characterization of science, Jesus was wrong to say ' I am the way.'

And to talk about "now" as a concept that applies throughout the universe is like trying to talk about "up" or "down" in space. Now really does not apply throughout the universe any more than up or down does. As a consequence, there really isn't a "before" or "after" for events throughout the universe. That should give you a clue about the main assertion of conscious life after death.

Let's go back to our example of Andromeda. If we were regarding the word "now" as including Andromeda and us, then everything that happens between these two million years is neither past nor future with respect to both of us. If aliens from Andromeda wanted to visit us it would make no sense to say that their spacecraft is leaving now. The first light signal we get means that the "now" that we speak of is in the past. Whereas 2.14 p.m. our time is in their future. Similarly, if we sent light signals to Andromeda, our "now" is in their past so to include both of us in the word "now" means that the word "now" would be two million years old. That's an awfully long time to keep the past and the future at bay.[23]

14

Conventions Versus Prescriptive Vocabulary

When people started to use words like "reality," "time," even "epistemology" and "light," they simply didn't understand experience as we do now. And they certainly didn't understand how, in order to make sense of these terms, we have to distort what we actually experience. Experience itself, in other words, is a kind of a distortion in the numerous ways that we have seen. And because quantum theory and general relativity don't really agree in how they see reality, that's led to strains in vocabulary as well – for example, two very different characteristics of what we call light. We may have to change definitions of "time," "truth," "existence" and "universe" and what is meant by "cosmology" and "epistemology" in order to understand what we're talking about here.

For example, we are used to using these terms as if there was only one direction to time's arrow. When we talk about the thought experiment, there may well be two senses of existence that we have to worry about, several senses of reality and so forth. Can we do this? Can we take this seriously? Have we got it in us to be relativist enough to talk about our reality or "a" reality rather than "the" reality? It's a big step and perhaps a dangerous one to be this prescriptive about vocabulary. But it seems we have no other choice if we are to expand our consciousness in this way.

Some of this new usage may even be speculative. We're not in a position to manipulate the geodesics of light in the way we would have to in order to observe some effects, like different directions of time's arrow or what the event horizons of black holes can do to the speed of light. We may never be in a position to do this. In which case our vocabulary is based on speculations and thought experiments.

If, for example, we were to have a hard look at what we heretofore have called the universe in terms of both directions of time's arrow you might find the term pluriverse suits this expanded consciousness of epistemology better than universe. A pluriverse could be defined as one specific direction of time's arrow, let's say past to future. Whereas another pluriverse could be one in which time's arrow is going from future to past. There are other ways to talk about different pluriverses. We could talk about pluriverses in terms of different perceivers or even in terms of different decisions by the same perceiver. Or, for example, we could incorporate black holes as the beginnings of pluriverses, as though there were not just the Big Bang but a Big Bang of a specific pluriverse.

None of these ways of defining what we might come to mean as pluriverse have yet been conventionalised. In other words, we are using vocabulary that's not yet generally accepted.

You might say that I've been asking you to recognise that there are conventions or distortions in what we observe and yet I'm asking you also to step outside these conventions by using new vocabulary. Yes, it's a tall order. I'm asking you to do both of these things. But that's what an expansion of consciousness demands.

Some of these prescriptive uses dissipate when we talk about earth bound conditions, but that doesn't make them any less valid. It simply means that we're not so used to using them. For example, In the thought experiment in the next chapter there might not be enough room for differences in geodesics of light or curvature of light to make a difference in terms of the timing of light signals. The second set of time's arrows, where the future goes into the past, would disappear over shorter distances because the geodesic of light is not a factor. The distance is not big enough for a geodesic to operate enough to make much of a notable difference. But that doesn't make it any less valid than the time sequences that we're used to, let's say, using time where it moves from before to after. It's simply that we're more used to talking about the before and after variety of time sequences. For example, if we move towards a distant light source, the second type of time's arrow from future to past would dissipate. But we'll talk about that in a minute.

All this begs the question of whether we can really expand epistemology beyond the traditional limits of what we see; what we experience. In other words, can we somehow go beyond the conventions necessary to observe what we've outlined?

We're already doing this by proposing to introduce a thought experiment, a hypothesis. If we treat this prescriptive use of vocabulary as part and parcel of that hypothesis then, on those grounds, perhaps we can be allowed to be this prescriptive about vocabulary. It seems that we have to be in order to accommodate the expansion of consciousness that we're going through in this adventure.

Can we say that we know something without actually seeing it and experiencing it? Think of the crunch of gravel in your driveway at 5:00 p.m. You're used to your wife returning home from work at this time. On this occasion, you haven't seen her but you're making an inductive assumption that it's her when you hear the sound of the gravel. If you say you know it's her, then you're inadvertently increasing your definition of epistemology to what's a little bit more speculative. In other words, this weakened sense of knowing is included as epistemology. But should it be? I mean, this is basically what we're doing in this intellectual adventure. We are, on the basis of a thought experiment that we're about to introduce, asserting that there is conscious life after death.

Your belief system, if you take these remarks

seriously and you believe them, should now be pretty topsy-turvy. You would no longer be an 'ersatz modal realist', that is you wouldn't believe that this is the only world or this is the only universe.[24] There are consequences of that, particularly if you feel that this is the only world or universe that you're living in. The other universes – pluriverses as we call them – are purely speculative. But even if you think that and believe that, there is a difference. You have expanded your consciousness to quite a large degree even to accept the idea of pluriverses.

15

The Thought Experiment

How can we assert that there are many pluriverses, that we can decide the size of them and that there is conscious life after death? How can we make these seemingly absurd statements?

Remember that we said you cannot travel outside the farthest star because light travels in curves or geodesics and turns in upon itself and, since light is the messenger of the universe (that is, you cannot observe or experience anything that isn't shown by light), any message it does not bring to you is outside experience and therefore outside epistemology. So the sky is not infinite in that sense.

Now, here's the thought experiment on which our assertion that there is conscious life after death is based.

Suppose we could deliberately manipulate light signals that we send out and somebody on a distant light source could do the same with light signals they send out. Imagine somebody on a distant light source sending two light rays – A and B – to Earth, one after the other. They send light ray A around the event horizon of a black hole several times where it doesn't fall in but it is delayed. Then they send light ray B, which left after A but does not go around the event horizon of the black hole. It takes a more direct route. In this case light ray B would arrive before A to the observers on the Earth even though it left the distant light source later. They are using the black hole to deliberately delay the first light signal by manipulating the geodesic of light ray A as opposed to light ray B. Now, let's say that we're doing the same thing to light rays that we're sending to people on the distant light source.

If this process is repeated continuously, our observations would involve time reversal, that is we would see a continuous stream of light signal B's before light signal A's, even though the stream of B's were sent after the stream of A's on the distant light source. If we are distant enough from the light source to allow for these manipulations of light signals, then time reversal is a very real prospect. This means that time has a second arrow, if you like, that starts from what we call the future and progresses to the past.

In terms of the way we experience both sets of observations, one engendering time reversal and one not, they are both real. Suppose we could include both sets of time signals that we'd arranged as one observation. You could see time progressing both ways in one observation. Or suppose we took a film of an earlier set of light signals (say T1 "before" progressing to "after") and compared them with a film of the later set of light signals (T2 "after" progressing to "before"). We could talk about time reversal as "a" reality (i.e. involving T2) and not "the" reality.

Again, suppose you are communicating with someone who is travelling towards the light source and you are not, and you are far enough away to apply both directions of time's arrow. You could get contradictory statements. You could say that so-and-so is both getting older and younger, which is nonsense. So what you have to do in order to make the truth conditions fit the situation we're in here is to make it clear that the truth conditions belong to only one direction of time's arrow. So you would say that, in terms of past to future, the person you're observing is getting older, but in terms of future to past, he's getting younger. Without these built-in conditions: yet another layer of theory-ladenness, you'd be talking nonsense. How do you like the idea of appearing to get younger in the sight of other people?

So theory-ladenness strains vocabulary, even making it prescriptive enough to refer to things that we're not used to referring to. For example, it might be that we would like to use a reality or pluriverse in terms of one direction of time's arrow and another reality in terms of another direction. But this is all up for grabs at the moment. This is all so new that no-one has come up with any conditions as to how it might work. We're now experiencing time in a new way, in a way scientists thought was not possible.

If we were to manipulate these light signals so that we fulfilled the notion of finiteness (that is, we travelled in a rocket powerful enough to keep up with a light signal until it came back to where it started) and we were able to manipulate the light signal so that we could send it around the event horizon of black holes if we wished, we could presumably control the size of the pluriverse that we observe. For example, if we sent a light signal around the event horizon of a number of black holes several times, then it would take longer for the light to come back to us than light sent along a different route without black holes. The pluriverse of the first light ray that went around the black holes might seem bigger because the light seemed to travel for a longer time than a pluriverse imagined using the second light ray even though it might actually be smaller.

This has a dual purpose of getting rid of two epistemological anomalies, one being whether the universe expands forever and the other where it contracts towards a big crunch. Both of these are epistemological problems. We can't visualise infinity in either of these senses.

So our manipulation of the size of our pluriverse gets rid of these problems for our theory of knowledge. Epistemology, as we now see it, would be prone to subjective constraints. We would control what we know. We'd be godlike in that respect. What we know would become subjective in this sense. That's a blow to what we normally mean by epistemology. A low blow, you might think. But if we're to follow where reason leads us, then it makes very good sense to say this, even though it seems very odd and maybe even counterintuitive to how we normally think.

Truth would take on this subjective tinge as well. We would be able to control and manipulate what we know is true, and that's a rather frightening thought. Reality, again, would be what we make it. There wouldn't be much difference between, let's say, evaluative and scientific language. The only distinction would be that scientific language would involve more of an effort to define by measurement and numeral strictures, what we are talking about. The constraints are really the conventions we have used in order to talk about observable experience. Within those constraints, we have a freedom that we didn't heretofore realise we had. The constraints, or determination, if you like, in the freedom determination argument, would be conventionalised by us – that is, the constraints are put there by us in order to understand. They are necessary to us in that sense.

One consequence of all this is that it would seem that if there is dark matter and it does have a gravitational effect on light, then some of the universe, or more accurately, our pluriverse could if that's so, (and it's a mighty big 'if') be getting younger. In other words, there's no overall composite time sequence that leads from the Big Bang or a Big Bang to infinite expansion or the big crunch. Pluriverses in any case would render that sort of talk obsolete. There would be no Big Bang. Conversely if there is no dark matter there is no Big Bang on that account either. In the literature, it is said that the Big Bang depends on the presence of dark matter.

All of this is hypothetical, of course. We can't manipulate light physically to this extent. All we can do is imagine it in our thought experiment. But suppose we could chop up these light signals and they pertain to another person's life. We could look at stages of that life at will. We could see as much of that life as we want to and the other person could see as much of ours. Rather like looking at old films. Or holograms. If we could chop up light like this, we could look at several stages at once or see someone coming and going at the same time.

And then I thought, "Sod this," and started ageing backwards.

16

Existence and Conscious Life After Death

What does this, after all, do to the concept of existence? We could say that the present use of the word "existence" is stretched beyond its remit here. For one thing, we're seeing a person's life that could be quite apart from the direction of time's arrow that we're used to, the one that involves before and after, in that order. We could say that existence as it's now seen splits into two. Existence involving the second and third person, the pronouns "he" and "you," would involve both directions of time's arrow. And existence involving "I" would involve only one, the one we're used to. These are strange outcomes but they're outcomes as direct implications of general relativity – that is, they are where reason leads us.

This second aspect of our thought experiment begs the question of existence. If we are still using the definition of universe as "all existing things."

Then what are all existing things? Are they the same for all of us? For a thing to exist, for example, do we have to be intersubjectively aware of it? You may be aware of parts of your pluriverse that I am not aware of and vice versa. You could be aware of light signals that I am not aware of and vice versa. We have to figure out ways of agreeing on what exists: call it intersubjective agreement.

Again, this sort of talk seems to render the notion of the Big Bang as the beginning of the universe obsolete. We could see hypothetically as much or as little of a life as we wanted to. According to time reversal, we could see a person in his coffin and, at the same time, see him running a mile or drinking a beer or getting married. This means that, in the second and third person tense of a language, we would be able to talk about conscious life after death. I, in the first person, have consciousness of your, his, or her life after your, his, or her death.

Unfortunately, however, we cannot do that with our own lives. Nevertheless, this would add an extra dimension to life as a whole. We seek a kind of immortality. We want to be remembered after we are dead and these hypothetical manipulations of light signals do just that. So, if existence is split into two in this way, we can see that the second part involves our assertion directly. There is conscious life after death. Again, the comparison with watching old films with actors who are now dead is a good metaphor for this. According to the first person though. We are not conscious of our own life after our death. So here we answer the conundrum posed in the beginning of this book: how there is and is not conscious life after death.

So, to review, you could purposely make your pluriverse larger or, conversely, you could avoid as much gravitational matter as possible and make your pluriverse seem smaller. I could do the same. The point here is that we have a personal choice in making the pluriverse the size we want it to be. Our pluriverse has to be big enough for us to be able to manipulate the geodesics of light as we want. Within these limits we can control the size of our pluriverse. To that extent, we're all little gods. Empiricism then involves our efforts to turn personal measurements and observations into intersubjective ones. Epistemology then becomes the effort to co-ordinate our different observations into what we could call intersubjective measurements of those observations so that they become communicable, measurable, comprehensive and understandable. The same is true of all the terms we use, however prescriptively, in a "new" way. Like the mountain, like the piano, we simply have to find ways of using words such as existence, consciousness, identity, time, space and pluriverse, so that they make sense to others, even if we have to offer up theory-laden measurements or accounts of "partial" holism (whatever that is) in some cases, or "new" definitions in others.

I'm sure in some small way his spirit lives on ....

17

Existence and the Third Arrow of Time

The truth conditions of our existence could be changed by the observations of others, supposing they manipulate them. All this is, again, hypothetical of course. We can't actually do this. No one truth table or reality is absolute. There are many that might even be regarded as incommensurate with each other; that is, they don't have much meaning to each other. As a consequence, those who regard truth tables as the final arbiter of meaning could be mistaken if those truth tables are presented to us in the traditional sense.

So to exist is to exist as one of an indefinite number of observable beings. And this indefinite number can be changed. This is a scientific implication of general relativity and would back up David Lewis's insistence on the reality of a plurality of worlds, as he puts it. It would be an argument against our ersatz modal realism or the suggestion that this is the only real world and that the universe that we live in is the only real universe.

As I said at the beginning of this intellectual adventure, these conclusions about what it is to exist and what it is to observe are so different from what we thought, that they are hard to take in, hard to believe. And yet this is where reason leads us. This means that we know only some facets of existence until we can hypothetically control the geodesics of light signals. These manipulations, once more, depend on great distances. Where these distances are minimised or reduced, the effect disappears and the truth conditions become subject to one direction of time's arrow only.

A few years ago, I would never have expected to be able to assert for good reasons that existence could be split up in terms of personal pronouns, or that we would end up with the term pluriverse rather than universe, or that we could evaluate the pluriverse personally conceived in terms of size and create our own pluriverse like small gods, or that our experiences of each other could be determined so subjectively. But this is what this intellectual adventure has shown us. And now we have a third arrow of time to consider.

Suppose we were in a spaceship near the event horizon of a black hole and we turned our rockets up to full power so that we stayed in a position that was stable in relation to the black hole (we'd have to keep our engines firing pretty intensely to resist the enormous gravitational pull). The powerful gravity of the hole implies that time slows down for this rocket. If the rocket stays near enough to the horizon for one hour and then moves away, we would find that for those outside the rocket centuries have passed in that hour. The closer the rocket stays to the event horizon, the slower, with respect to the outside, time runs for it. So travelling to the past might be difficult but travelling to the future would be easy. We need only get close to a black hole with a spaceship, keep within its vicinity for a while, and then move away. So this is one way we could hypothetically extend our lives according to this thought experiment. That is, the "I" pronoun, the one that we're used to living with, would actually see the future once we moved away from the black hole.

On the event horizon itself, time stops. If we

got extremely close to it and then moved away after a few of our minutes, a million years might have elapsed in the rest of the universe (or our pluriverse).

Again, this begs the question: what are the limits of epistemology? Can it be expanded, for example, to consider some of these hypothetical uses of the words "pluriverse" and "time's arrow?" Can it depend, in effect, not just on empirical considerations (i.e. what we can see or experience) but also on the implications from what we can see?

We can get into arguments and discussions about what is actual or what exists. David Lewis considered that "actual" or "now" or "what is real" is very different for all of us and, to that extent, he believed we live in different worlds. He treated these different worlds as real. [25]

What we have learned about epistemology is that it depends very much on subjective experience that we then have to conventionalise – or distort, if you like – in order to make it intersubjective and understandable to others.

The strains we have put on vocabulary suggest an expanding consciousness. You may not be willing to have your consciousness expanded, that is up to you, but the offer is open. Again, all of this may or may not be correct. Implications are shakier than inductions and this assertion is an implication based on general relativity. I have tried to argue for these implications on good enough reasons based on the science that we know, but they may not be good enough reasons.

If, on the other hand, the implications are correct, then the results are nothing short of mind-boggling. For example, talk of the difference between a pluriverse as everything we can observe and the universe as all there is, was not even considered in Einstein's day, only one hundred years ago. David Lewis did not consider putting his speaker and observers on Andromeda. Perhaps Penrose, in saying that the universe can be explained mathematically, is not even considering the possibility of all the theory-ladenness or distortions that we have been describing and that he is, in fact, applying mathematics to a specific theory in a specific language rather than to anything that he thinks of as the universe itself. And that's a very different perspective from the one that he and Hawking seem to think they're involved with.

In order for the statements made by science

to be correct or incorrect, it has to be more conventionalised than we had previously thought. The main theses of this little book have now been presented. Two facts have emerged from this mental adventure: (i) we have to be far more theory-laden or conventionalised, to experience in the way that we do. How these conventions actually work in relation to what they are expressing is anybody's guess. (ii) The vocabulary we normally use seems inadequate for the sort of assertions made here. This could suggest an expanding consciousness which may be very exciting for you. Both of these supposed facts challenge epistemology as a certainty. This means redefinitions on a large scale. In addition there is a partial explanation of what it is to be conscious (cf. chapter 1).

Careful, time appears to stop
in this region of the lounge.

18

The Conventional Nature of $E=mc^2$

We now come to something so controversial that it's hard to present it to you. Nevertheless, I think I must. It is in fact a third major consequence of the thought experiment and of all the conventions that have preceded it in this little book. It has to do with our ability to control the size of our pluriverse. I want to challenge the efficacy or the fundamental nature of the famous equation $E=mc^2$. Now it seems absurd to do so because the use of the equation has produced practical results, however horrendous, like H-bombs and A-bombs. But in this chapter we are applying the equation in a different context and we shall see whether it survives this very different application.

First, we need to see not only how the speed of light ("c" in the equation $E = mc^2$) needs these conventions and distortions that we've talked about but, in fact, how it is itself a product of them and therefore not an exception, not fundamental to our understanding in any unique way. This despite the fact that it is fundamental to our understanding of special and general relativity.

As we have already seen the greater the curvature of light or geodesic in its journey, the longer the light will take to travel between the light source and the observer. This means that we have to build in conditions in order to talk about the speed of light as a constant. But how do we do this? The constant value of the speed of light is itself the factor by which we calculate the distance of the light source. Again, if there is any delay due to the amount of geodesic or curvature in the signal, c as the constant speed of light, is a posit for this result. It's what we use to measure it. And yet we have to find the amount of this delay in order to build c back up into a constant.

If the light source is moving towards us or away from us, c as a posit is necessary as a factor in this information. So the problem generally is that the information required to establish c needs c as a posit or a factor. And this sets up a circular argument, which won't do. If we have different values for light signals, as in the above example where two or more light signals come in different geodesics or curvatures, these sources of information would have to be obtained to prop up *c*, otherwise we would simply talk, as the above writer Clifford Will writes (see note 13), about the slowing down of light.

So, in effect, without these built-in conditions, *c* may have a different value for different observers. All this suggests another form of theory-ladenness involving $E=mc^2$. This theory-ladenness itself has to be posited if we are to still regard c as a constant. In other words, we have a good deal of propping up to do. But then it would appear that $E=mc^2$ isn't an equation that's basic to observation. It's an equation that we make by conventionalising our observations.

This is odd to say, in a way, because experiments have shown that a light signal always leaves an object at a constant speed, 186,292 miles per second. But this result is presumably from measuring one light signal at a time in observations over short distances where delays due to the bending of light is not a factor. If, however, two or more light signals travelling over great distances are included as part of the same observation, doubts about the constant speed of light creep in.

Suppose your eye were a million miles wide and you saw two light signals from the same source arriving at different times. In order to posit *c*, the constant speed of light, you'd have to already know the distance of the light source from your eye at the instance of dispatch of each signal. Suppose, in fact, you were travelling away from the light source at the speed of a radium electron, very fast , say, about 70,000 miles per second. But, first, how do you know the distance of the light source without positing *c*? Secondly, suppose as in the first thought experiment one signal is more delayed than another. Again, you have to posit c in order to determine this. Also, the actual distance as a combination of one and two needs the posit *c*. Furthermore, you have to know whether the light source is moving towards or away from you. All this information requires *c* as a posit.

The problem is that the information required to assert that *c* is a constant speed can itself only be obtained by positing *c*. This circularity is self-defeating as an argument for the validity of c as a constant. All this information would be necessary supposing you've got different values for *c* and supposing you could otherwise identify the source of *c* as the same event. None of this is apparent in the sort of experiment conducted over short distances with one light signal at a time, to show that light leaves a body at a constant speed of 186,292 miles per second.

Zhu Yin states that "in order for the…value of *c* to be measured, the times of emission and reception of a light signal must be exactly known but, according to special relativity, those times cannot be known unless the value of *c* is known."[26] This is a circular argument in philosophical terms, and therefore it is an invalid argument for *c* as a constant.

What I'm saying is that you can't at the same time use a factor such as *c* as a constant speed of light as part of a posit in order to get the answer that is the value of *c*. The constant can't occupy both roles at once in any kind of a valid argument.

So, what emerges from this second thought experiment is that *c* seems to be an idea propped up by calculations and not an observation, however laden with theory, on which we base our knowledge of the universe or more accurately our pluriverse. The possibility of a variable speed of light challenges the Doppler effect and, as a result, the Hubble Constant.[28] The Hubble Constant suggests that redshift[29] is not random but is directly proportional to a galaxy's distance from us. The apparent curved space in which light travels and, consequently, the apparent distance could be observed as altered for different observers supposing that they were in vastly different parts of the universe relative to the observation. Therefore the amount of redshift would be changed for each observer. There would be no one value obtainable intersubjectively. Consequently the entire universe, if we wish to revert to that sort of vocabulary, does not present us with a definite age.

The Hubble constant dissolves even for one observer since two or more values for the redshift could be obtained from two or more light signals emanating from the same source. (There is a vast and intricately argued literature on the topic, if you are interested in the philosophical problems of identity involved here.) Furthermore, there would be no actual position and speed of light that is true for all observers any more than there is a fixed position and speed for a quantum of light. To this extent, the uncertainty principle might become a facet of general relativity. $E=mosl^2$, where osl is the observed speed of light, might be a better equation according to this total relativity than $E=mc^2$. Osl is less theory-laden and therefore less distorted than $E=mc^2$.

The curvature of light implies a varying speed for different widely spaced observers or for a single observer of several different light signals who sees what may appear to be "the same event arriving by different routes." This would mean that c and therefore E had different values for different observers or from one observer supposing one light signal was "slower" than another.

If there are measurable distinctions between two or more light signals leaving a source, however minute due to the curvature of their geodesic, then c as a constant speed of light remains an abstraction of special relativity, particularly if this instance of the light signals is generalised. The constant speed of light would be an ideal never quite realised in nature. I doubt the equation will ever be put aside. It works too well in the contexts in which it is used. But according to this discussion it should not be regarded as the fundamental equation of the universe. It would be more natural to talk about the different speeds of light rather than its slowing down relative to some ideal constant that does not apparently exist in our observations.

Will that be all sir?

19

A Summing Up

According to the rather chequered history of metaphysics, neither of the following implications (i) time reversal and, [ii] the relative speed of light, can be established as a scientific fact. Nevertheless, there are good scientific reasons for supposing they are true. So talk of the universe and the Big Bang now seem obsolete. Elsewhere I have argued against the Big Bang as an established fact.[27] There seem too many good reasons in the literature against it. In any case the idea of the Big Bang as a first cause challenges the whole idea of cause and effect. There is no effect of a first cause.[28]

As far as I know it hasn't been pointed out (forgive me if I am wrong) that conscious life after death can be posited using just one dimension of time, the only one we are used to: before and after. So it could be argued there is no need for the above thought experiments. However since we have conventionalised our experience enough into objects to invent time, time reversal and even the stoppage or near stoppage of it, and if we take the event horizons of black holes seriously this would add one and maybe two extra dimensions to this assertion. And this could be regarded if it is right, as expanding our consciousness in this direction. Without objects there is no up or down or before and after, and therefore no real consciousness of time or of space. Danker L. Vink says "In the universe as a whole the two directions of time are indistinguishable just as in space there is no up or down."[30] So we use "objects" as we have conventionalised them to suggest that conscious life after death operates in all three directions of time's arrow, and not just one. So the fact that we can conventionalise time into three dimensions rather than just one, expands our consciousness of life after death.

In conclusion, when you make a scientific statement you are evaluating just as much as when you say something is good or bad. The difference being that value statements about good or bad seem more personal than scientific statements are about what's true or false. But in order to get to what's true or false we have to find a way of coordinating our subjective experiences into something that's intersubjective and that can be describable and communicable and understandable. That is where all our distortions and conventions come in. We don't do this quite as much with value judgments, hence scientific statements look different from evaluative ones. They look as though they are more objective.

Epistemology, the science of knowing, may now look somewhat different than it did at the beginning of this little intellectual adventure. Also, the distinction between scientific statements and value judgments is rather different than we had supposed. And who would have thought there are pluriverses? Or many worlds? Or that we could split up existence in the ways that we have? Or decide the size of a pluriverse? "There are more things in heaven and earth, Horatio, than are dreamt of in your philosophy." I submit that we don't need religion to tell us this. Science does a pretty good job, as we have just seen. But no longer can epistemology be certain in the sense that there is objective subject matter. We have to make it objective or intersubjective at least. Even the metaphorical image of a triangle of cosmology or physics we make ourselves. This means that mathematical equations may be applied subjectively in the first instance and made intersubjective by means of various conventions. A lot of these conventions involve articulating equations. That is how mathematics helps us to objectify experience so that we can understand, work with and change each other's perspectives and understanding.

The result of all this is a new way of looking at science. What we know is irredeemably subjective in a referential sense. We make observations into experimental data through mathematics and measurements, all of which are conventional, man-made, and not in nature, and therefore they are a prismatic distortion of experience itself.

These issues of theory-ladenness or distortion reign across all subjects and subject matters and, consequently, each different subject, as it considers the issues involved, considers them from a very partial, or what we might call biased, angle. Even so, human consciousness can be raised up and expanded in terms of these distortions and this is very exciting.

One final point to make is that existence in terms of epistemology may vary from observer to observer. Hearsay could be right in positing existence, whether yours or mine. Again, our existence would depend not just on observation but on somehow conventionalising it enough so that it's understood by others and perhaps could be observable by others. It is within such conventions or confines that we could be said to exist consciously after our own death and even, in some cases, be getting younger. But we are dependent on each other in ways that we would never have guessed at the beginning of this intellectual and, if I may say so, spiritual adventure.

Thank you for joining me in this adventure. I hope you enjoyed it.

Notes

1. Stephen Hawking, *A Brief History of Time* (London: Bantam Press, 1988), 11.
2. William Shakespeare, *Hamlet,* Act 5, Scene 1.
3. Albert Einstein, *Quotable Einstein* (Princeton: Princeton University Press, 1996), 169.
4. Ibid, 175.
5. This quote appears in Albert Einstein, "Physics and Reality," *Franklin Institute Journal 221*, No. 3, March 1936, (reprinted in Out of My Later Years), 349–382.
6. Roger Penrose. This quotation is to be found as the first one in the quotes of Roger Penrose online. It is supposedly in his book *The Road to Reality* (New York: Knopf, 2004), but I have to say that, after much searching, I couldn't find it there.
7. Stephen Hawking, *A Brief History of Time* (London: Bantam,1988), 175.
8. Willard Van Orman Quine, *Word and Object*, (Cambridge, Mass.: MIT Press, 1960), 73–78.
9. See in particular Wiggins, David, *Identity and Spatio-Temporal Continuity* (Oxford: Oxford University Press, 1967) and Saul Kripke *Identity and Necessity and Identity and Spatio-Temporal Continuity* (Oxford: Blackwell,1967).

10. Stephen Hawking, *A Brief History of Time* (London: Bantam,1988), 124.
11. *The Oxford Compact English Dictionary* (Oxford: Oxford University Press, 1996), 1138.
12. Carlo Rovelli, *Reality Is Not What It Seems* (London: Allen Lane, 2016), 78.
13. Clifford M. Will, *Was Einstein Right?* (New York: Basic Books,1986), 13.
14. Jaroslaw Mrozek, *Filozofia Nauki*, 4, 2001, see abstract.
15. Ibid.
16. William M. Honig, "Mathematics in Physical Science or Why the Tail Wags the Dog," *Physics Essays*, 13, No 4, 2002, 518–519.
17. W.B. Yeats, "Sailing to Byzantium," *The Collected Poems of W.B. Yeats*, (London: Macmillan, 1933), 191.
18. See Alexander Matthews, *Physics Essays*, 18, No. 4, 2005, 462–67.
19. 'Event horizon' can be defined as the border of a black hole, that is an area where the very edge of life as we know it takes place. Beyond that the electrons of all atoms would be stripped bare and the density would be so great that present levels of mathematics and physics cannot be practised within a black hole. Events such as we know them would cease to exist. So if light is trapped going around the event horizon of a black hole, it is being kept by the gravitational pull of that black hole but not near enough for it to get sucked in beyond the horizon where events take place.
20. Stephen Hawking, *Brief History of Time*, (London: Bantam, 1988), 175.

21. Carlo Rovelli, *Reality Is Not What It Seems*, (London: Allen Lane, 2016) 229–231.
22. Ibid.
23. The late philosopher David Lewis based the practical reality of his plurality of worlds, worlds that he really believed in as facts, on the supposition that our interpretations of 'now' and 'actual' were very personal and idiosyncratic. Our world views, therefore, were different. He could have reinforced this assertion had he placed some of his observers on Andromeda. See David Lewis, *On the Plurality of Worlds*, (Oxford: Basil Blackwell, 1986), particularly Chapter 2 "Paradox In Paradise?"
24. David Lewis, *On the Plurality of Worlds*, (Oxford: Basil Blackwell,1986), Chapter 3.
25. Ibid.
26. Zhu Yin, "Investigation of Special Relativity," *Philosophical Essays*, Vol. 15, 2002, 364.
27. Alexander Matthews, "The Universe Has No Beginning? Doubts About The Big Bang Theory," *Physics Essays*, Vol. 18, No. 4, December 2005, 462–466.
28. Ibid.
29. Redshift means that the light develops longer wavelengths as an object which is shining that light moves away from us. This is just as the sound of a train whistle drops as the train moves away. Red waves have a greater length and these waves and light takes on the guise of a greater length as it moves away. So there is a redshift in the light of objects which are getting further away from us.

30. Danker L. Vink, *Physics Essays*, Vol. 15, 2002, 405. In deep space if there are no coordinates, there is no time. So this may be one way of talking about eternity. Also, entropy clashes with the idea of a regenerating pluriverse, since there is a run-down or loss of energy. But entropy itself raises questiions, that is, how did it get the energy in the first place?

Bibliography

Einstein, Albert. "Physics and Reality." *Franklin Institute Journal* 221, No. 3, March 1936.

Einstein, Albert. *Quotable Einstein*. Princeton, NJ: Princeton University Press, 1996.

Einstein, Albert. *Relativity: The Special and the General Theory*. New York: Wings Books, 1961.

Gleick, James. *Genius: Richard Feynman and Modern Physics*. New York: Little Brown, 1992.

Hawking, Stephen. *A Brief History of Time*. New York: Bantam, 1988.

Honig, William. "Mathematics in Physical Science or Why the Tail Wags the Dog." *Physics Essays*, 13, No 4, 2002.

Kripke, Saul, and Wiggins, David. *Identity and Necessity and Identity and Spatio-Temporal Continuity*. Oxford: Basil Blackwell, 1967.

Kripke, Saul, and Wiggins, David. *Identity and Spatio-Temporal Continuity*. Oxford: Basil Blackwell, 1967.

Lewis, David. *On the Plurality of Worlds*. Oxford: Basil Blackwell,1986.

Matthews, Alexander. *Physics Essays*. Vol. 18, No. 4, December 2005.

Mrożek, Jaroslaw. *Filozofia Nauki*, 4. 2001.

Penrose, Roger. *The Road to Reality*. New York: Knopf, 2004.

Quine, Willard Van Orman. *Word and Object*. Cambridge, Massachusetts: MIT Press, 1960.

Rovelli, Carlo. *Reality is Not What it Seems*. London: Allen Lane, 2016.

Vink, Danker. *Physics Essays*, Vol. 15, 2002.

Will, Clifford. Was Einstein Right. New York: Basic Books, 1986.

Wiggins, David, *Identity and Spatio-Temporal Continuity*. Oxford: Oxford University Press, 1967.

Wittgenstein, Ludwig. *Philosophical Investigations*. Oxford: Basil Blackwell, 1978.

Yin, Zhu. "Investigation of Special Relativity," *Philosophical Essays* Vol. 15, 2002.

Index

Born in New York City in 1942, Alexander Matthews taught philosophy at a number of universities between 1975 and 1989. In 1986 he was awarded a Visiting Fellowship to Princeton University.